South Korea Tours of Duty

AF879734

Vickey Joins the Military Series

Book 6

South Korea Tours of Duty

Book Six

Table of Contents

First South Korea Tour of Duty

Kunsan Air Force Base

The time was slowly arriving for me to leave my second duty station, White Sands Missile Range, but I still hadn't received my next orders. As I waited for those orders to arrive, I continued with my duties, testing new weapon system equipment. Finally, the orders came through, sending me to another far-off location; I was off to Asia, my third continent.

As news of my next assignment spread through the unit, I received many condolences, which was bizarre. I was headed to Korea, a place that apparently nobody wanted to be assigned. The stories, and there were many, would come from individuals who, themselves, had not ever been to Korea, but they had no problem offering up stories and opinions of the duty and the far-off country.

I'd always preferred to form my own opinions based on personal experience rather than relying on others' views. So, I approached my assignment in Korea with an open mind. Besides, it was only a one-year tour of duty. I figured I could handle anything for that long.

The journey to Korea was going to be long, but I felt better prepared to handle the jetlag that time. I had dealt with it before during my first overseas assignment, but back then, I hadn't recognized what I was experiencing until days later.

I arrived in Korea in the summer of 1995 and was immediately greeted by the unique aroma of the countryside. Korea, a 750-mile-long peninsula on the eastern part of the Asian continent, was where I would spend my tour of duty. South Korea, with its many mountains, is bordered by the Yellow Sea and filled with regions of forests and paddy fields, which contributed to the welcoming scent upon my arrival.

When I arrived at my new duty station as a mid-level noncommissioned officer, I was expected to take on a team leader role. However, first things first, I needed to in-process into the unit. During that process, it became clear that there were no living quarters available for a female noncommissioned officer, which presented a conundrum.

An unused trailer next to the barracks, where the soldiers lived, became my quarters for the next year. It was equipped with its own bathroom facilities, which made it ideal for living. I spent my first few weeks at the new duty station turning that trailer into a comfortable home.

The duty in Korea involved the same routine as previous assignments—extensive training, field exercises, and making the most of our down time. Since there were only a few Army soldiers stationed at the Air Force base, we quickly became a close-knit working group, almost like an extended family. We became fast friends and spent a lot of time socializing together when we were off duty.

Midway through my tour, our platoon sergeant, our first-line leader, had to leave the country for personal reasons, and I stepped up to assume his position. I volunteered eagerly for the responsibility because it was a new role for me and it would enhance my record. Holding a position not commonly occupied by someone of my rank would give me an advantage during promotion evaluations.

In my role as platoon sergeant, my responsibility was ensuring the well-fare of the platoon personnel, about 40 to 50 soldiers. That encompassed overseeing their training, accountability, and overall care. The job was easy because we in the platoon were friends, and we were keen on helping one another out. So, the transition from being just one of the platoon members to leading the platoon was smooth.

Shortly after assuming the role of platoon sergeant, tensions between North and South Korea escalated, catapulting our mission from routine to a state of heightened alert. We remained on the missile site for days, only returning to the rear every other day, only to shower. That period put our training and capabilities to the test, a scenario not uncommon in our experiences in Korea.

Military forces frequently engage in field exercises to assess their capacity to execute missions as directed by leaders, evaluating both their preparedness and the capability of allied forces to fulfill their designated roles. That practice is no exception in Korea. However, what set the exercises in Korea apart was the involvement of millions of soldiers from various nations operating in close proximity to one another. Those exercises invariably coincide with heightened tensions, underscoring the complex dynamics at play in the region.

The two halves of Korea always stood ready to attack or defend the limited, mountainous terrain. The two armies, North and

South Korean, had very distinct and different fighting capabilities and any conflict between the two had the potential of killing hundreds or even thousands of individuals during the initial stages of a conflict.

Therefore, when our small air defense unit was assigned the responsibility of operating our equipment around the clock, we understood the seriousness of the situation. Exercises always seemed on the verge of becoming real in Korea, so it was always critical that the equipment remained operational in case any exercise became an actual conflict.

We didn't doubt the equipment, after all, it had been tested in the Gulf War conflict, so we knew its capabilities. We just had to maintain the equipment, maintain our skills, and remain in a state of readiness.

Throughout numerous exercises, my platoon consistently demonstrated flawless performance. Our maintenance crews promptly restored any equipment that experienced down time, ensuring minimal disruption. The urgency to keep the weapon system up and running was critical during a constant state of heightened tensions. Our ability to maintain the weapon system and keep it operational during twenty-four-hour operations caught the eye of our Warrant Officer.

Our Warrant Officer, the technical expert for our weapon system, praised our operational abilities, notably during challenging monsoon and winter operations. It was during those instances that I suspect he made the decision that I should consider applying to the Warrant Officer Candidate School. Although he didn't mention it at the time, it wasn't until months later, when we found ourselves stationed at the same duty station that he broached the topic, a topic that he would become utterly determined to get me focused on, making numerous pleas in hopes that I would apply to the elite group.

But while still in Korea, the relentless cycle of field exercises persisted, and the Korean climate, whether in summer or winter, proved to be unforgiving for both personnel and equipment alike. In the southern part of Korea, the climate was humid, and it was affected by monsoons, heavy rains during the summers and frizzed temperatures during the winter. That made for some exceedingly difficult field exercises with damage to equipment, the washing away of field living tents, and cases of cold injuries. Field duty in Korea during the summer or winter was, many times, quite brutal.

Recovering after saving the tent from monsoon rains and winds!

The winters in Korea were exceptionally harsh, surpassing even the cold I experienced while stationed in Germany. Our work and field exercise sites, situated along the shores of the Yellow Sea, subjected us to biting winds and bone-chilling temperatures. Despite donning every piece of winter clothing issued by the Army, I still found myself inadequately protected against the relentless onslaught of the bitter cold.

Winters near the Yellow Sea!

Irrespective of the external conditions, our missions had to proceed without interruption. We had to learn to perform our duties under all conditions, and in many cases, with joint personnel during joint military exercises.

Throughout the tour, all joint exercises concluded smoothly without incident. However, there was one instance where the extreme weather conditions proved too formidable for our multi-ton, lethal equipment; it was simply not suited for the environment. Thankfully, such clashes between the equipment and the environment were rare occurrences during our tour of duty.

When our multi-ton equipment got stuck!

With the equipment temporarily out of commission and while awaiting maintenance support, the soldiers ingeniously entertained themselves with friendly competitions.

As my time in Korea drew to a close, I participated in my final field exercise with the team. My tour to Korea turned out to be nothing like the horror stories I was told before departing for duty. Korea turned out to be another location where I would find great travel opportunities and friendships that I maintained for the remainder of my military career.

Once again seizing the chance to explore while stationed abroad, I embarked on a journey to Jeju Island, a province of South Korea. Its breathtaking landscape surpassed anything I had ever seen outside of a postcard—it was truly magnificent. I also had the pleasure of visiting some of the most picturesque beaches, soaking in the beauty of the surroundings.

The work was demanding yet fulfilling, the countryside held a wealth of fascination, the cuisine was nothing short of extraordinary, and the bonds of friendship I formed endured long after my tour ended. The experiences that others voiced about Korea were the total opposite of my experience. I had a great tour!

The year went by fast, as I expected. And as the tour was winding down, the personal bonds were solidified as many of us learned that we would be headed back to the States, to the same duty location.

Oh, happy days!

We rejoiced when we learned the news. While we might not be assigned to the same unit, knowing that we would be stationed in the same location meant that we could sustain the bonds we had formed. The news just put the cherry on top of a great tour of duty, and we would have an opportunity to continue the friendships we had made.

Friendships would continue!

I took away from the tour a love of the country's food, a respect for space because the country had a lot of people, and much of the country's habitats seemed to be squeezed into small areas; and I took away a fascination with the Demilitarized Zone, what a fascinating place, where you could sit in one room and be in both North and South Korea.

I would depart from my first tour of duty in Korea, leaving behind the memories of brutal summers and winters, the desert-like landscapes, and the distinctive countryside aroma, as I headed back stateside to the home of Air Defense.

Goodbye Kunsan!

Second South Korea Tour of Duty

Osan Air Force Base

In the 90s, during my first tour to South Korea, the country seemed to transition to a more robust economy. Although I was not actually aware of that transition at the time, I would be able to realize the transition during my second tour of duty.

My second tour in Korea occurred in the 21st century, specifically in 2007. The country had experienced many changes; the countryside aroma that met me as I landed during my first tour wasn't as potent. There were more paved roads, the highways were improved and filled with a great many motor vehicles, and there seemed to be a change in the social atmosphere.

South Korea had developed significantly, and they were technologically advanced. The country made so many great strides in many areas, such as music, entertainment, and innovation. It was as though I had arrived in an entirely new and modern South Korea. The country had undergone a transformation and modernization since my first tour, which was remarkable.

During my second tour, I was stationed at an Air Force air base once again, but this time, as a Warrant Officer, my duties diverged significantly from my first stint in the country. With the second tour I was placed in the officer's living quarters, I had permission to drive, and I was placed in a more liaison role and not an operator role. I was the conduit between my unit, located on a different installation, and the Air Force air defense element.

When I learned of my new role, I was ecstatic. I was located on Osan Air Base. The base was completely modernized and had touches of Western culture, from restaurants to living facilities and entertainment. It had many of the amenities that would be offered on a stateside installation. I wasn't sure what I had done to deserve such good fortune, but I was elated and vowed to take full advantage of my new assignment.

Before I could settle into my new, plush digs, I had to get settled in and learn what my new responsibilities entailed.

As I became familiar with my new assignment, I learned that the combined exercises in South Korea were far ahead of where they were during the 90s. They were far more advanced; that readiness in the defense of South Korea and the continued deterrence of attacks had advanced just as the countryside had. It would also be revealed that North Korean forces had also made significant advancements in their capabilities.

There were a number of situations that changed how combined exercises were conducted, including the operational control of the forces and the ability of North Korea to mount a nuclear warhead on a delivery system able to reach the US. Those major changes required significant adjustments in training to reflect the new requirements for defending our interests. The advancements by both South and North Korea posed challenges that would require a huge effort and much dedication to overcome.

As the resident Army air defense expert, my responsibilities included coordinating joint training with Air Force and South Korean air defense experts. That collaboration ensured that all parties operated from the same playbook regarding air defense. As training and field exercises began, all forces encountered a learning curve due to multiple upgrades in weapon systems and enhanced lethality capabilities, which compounded the threats we faced.

As with my first tour, that year-long duty involved constant training and field exercises. However, that tour introduced an additional element: Navy air defense. Inspired by lessons from the Gulf War and efforts to minimize friendly fire, all branches began to collaborate more closely on air defense missions. That required our Army air defense crews not only to train and work with Korean and Air Force counterparts but also to engage with Navy air defense personnel. Consequently, we spent time working and training in the air defense operation centers of each service.

Joint training with peninsula air defense crews!

Being embedded with the Air Force and Korean air defense presented no issues, as we were all located on the same air base. However, training with the Navy posed more challenges; my tactical air defense crew and I had to make several trips stateside to train with Navy personnel.

We're in the Navy now!

My crew and I operated alongside our Navy counterparts in the operations center on Navy cruisers. We trained together to distinguish between threats and friendly forces.

The importance of that training was clear to us—it was crucial for ensuring effective communication among the joint forces. Without clear communication, the risk of disaster in the skies was high, making the training both lengthy and intense.

Long hours of training on the Navy cruiser!

But as with all duty, there was some down time, and we were able to enjoy learning a bit about Navy life and enjoying being stateside, if only for a few days at a time.

Happy to be stateside, if only briefly!

But before we got to enjoy the down time, we had to fine-tune our joint air defense tactics. During our ship training, we conducted continuous tactical air and missile defense operations. We focused strictly on air operations, identification of air threats, and communication. Those three elements were the key to success for any joint operation.

As our time with the Navy crew ended, the air defense joint forces had a better understanding of each other's defense capabilities and how each trained to achieve success in the air defense arena. The collaboration among the South Korean forces, Air Force, Navy, and Army air defense was not only remarkable but also extremely beneficial.

Final Navy joint training, The USS Chancellorsville!

Although training on the USS Chancellorsville marked our final exercise with the Navy, it was not the last training event of our tour. We had several more training exercises scheduled upon our return to the Korean peninsula.

When we returned, we were almost immediately thrust right into another exercise, but the training was solely Army personnel, different from previous exercises. A crucial component of readiness emphasized in that final exercise was maintenance. The mission assessed the manageability of integrating training and maintenance cycles, with the intent to establish predictable cycles that would enhance the readiness of personnel, equipment, and training.

The mission played out for 30 days, with all units training to transition from conducting continuous air defense operations to being ready to conduct continuous operations within one hour. The exercise was an intense month-long cycling through operational readiness evaluations, deep maintenance of all equipment, and training, testing all air defense crews on their skills and abilities to perform at the highest readiness if called upon to do so in an actual conflict.

I managed to survive the rigorous 30 days, and all crews were certified to perform their duties in a conflict. But after those stressful days, it was time for some much-needed rest and relaxation. I looked forward to one last opportunity to travel before leaving the country.

During my second tour of duty in Korea, I had the opportunity to travel to Japan, Hawaii, and Guam, the latter known as "Where America's Day Begins," which is the territory's motto.

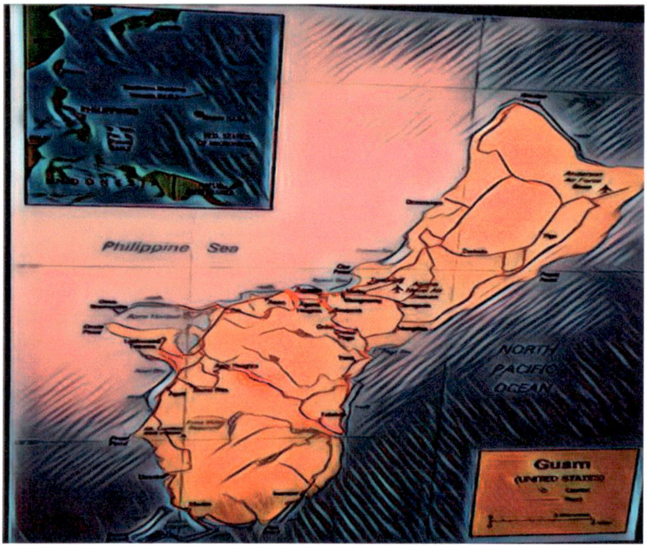

A small US territory!

Guam, a US island territory, is in the western Pacific Ocean and is distinguished by tropical beaches.

Beautiful waters!

The territory of Guam spans just over two hundred miles and can be driven end-to-end in about four hours roundtrip. Its economy is primarily supported by tourism and the US military, with the territory serving as a significant asset for military operations.

The territory was a beautiful vacation spot, and I was able to knock off one of the many US states and territories I had vowed to visit.

Shortly after my trip to Guam, I returned to my duties. As my unit's tour was concluding, we made sure to prepare detailed continuity documents. Those documents were essential to inform the personnel replacing us about the significant progress we had made in joint operations, ensuring a smooth transition and continued success.

My second tour in Korea, contrary to the negative experiences often expressed by others, was quite positive and left me with valuable experiences and friendships that I would cherish for years. While I can't speak for everyone, Korea was far from the worst assignment one could receive. If those who had negative views of their Korean tour, or those who never experienced it, had embraced the assignment more positively, there might not be as many horror stories about serving in Korea or the stigma associated with it. Both of my tours in Korea were rich with experiences, friendships, and opportunities for growth in my military career.

Made in the USA
Las Vegas, NV
07 November 2026

57675042R00017